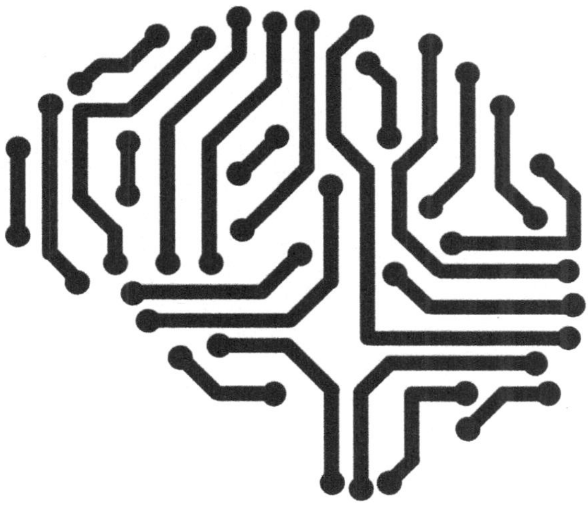

Published by Apttus Corporation,
1400 Fashion Island Blvd., San Mateo, CA, 94404. www.apttus.com

ISBN 978-0-692-85756-4

Text design and composition by Val Sherer, Personalized Publishing Services

Acknowledgements

We would like to thank and acknowledge all the amazing Apttus customers, partners and employees.

We would also like to thank those that helped make this book possible including Kirk Krappe, Ankur Ahluwalia, Ben Allen, Todd Browning, Man Chan, Wendy Close, Jeff Cowan, Geeta Deodhar, Michael Dunne, Amy Gardner, Derrick Herbst, Jason Loh, Daniel Louie, Brion Schweers, Sarah Van Caster, Patrick Wolf, Flora Wong, and Maria Pergolino.

Contents

1 Introduction

Quote-to-Cash, your company's revenue creation process, stands at the precipice of the next big wave to hit the enterprise—and this wave will be driven by artificial intelligence (AI) and machine learning. If you catch this wave, you are poised to capitalize on its competitive advantage and surge ahead of your rivals.

Today, AI is all around us and is increasingly playing a role in our daily lives. AI enables human-like cognitive capabilities and interactions, like chat and speech-to-text. Machine learning excels at pattern recognition to solve complex problems without the aid of explicit computer programming.

Whether it is Tesla's autopilot enabling a self-driving car, Apple's Siri or Microsoft's Cortana automated assistants, Google with its massive global data centers doing page rank and presenting search results, or refined suggestions on Amazon, Expedia or Netflix—machine learning is already enriching our daily lives.

Taking our cue from the consumer world, we're in the middle of a perfect storm for creating intelligent-first applications in the enterprise, as a result of ever-increasing piles of data, cheap storage and cloud computing power. By enabling these powerful technologies in the Quote-to-Cash process, we can make quoting, pricing, contracting, billing, fulfillment, and contract renewals more efficient and more valuable - which is nothing short of revolutionary.

So, where do you start?

Quote-to-Cash is, historically speaking, one of the most painstakingly manual and disjointed processes in every business. This set of cross-functional business process connects a customer's intent to buy and a company's realization of revenue, encompassing the entirety of sales, contract, and customer relationship lifecycles.

Quote-to-Cash is commonly one of the most poorly implemented processes, and often manually executed across an array of departments and platforms, with each step containing siloes of information which are not easily shared, leading to mistakes, gaps, and unintended outcomes.

Machine learning has the potential to knock down those siloes, not just by automating and streamlining the Quote-to-Cash process, but also by driving revenue through the maximization of business outcomes with data-driven intelligence.

Imagine if your sales team was automatically alerted to a new sales opportunity and had insights into the products and pricing that are most likely to win the deal. And, on top of that, they were provided guidance that streamlines approvals to close the deal faster. That is machine learning in action and it is available now.

In the following pages, we will explore these concepts, what it means to deploy machine learning in the most impactful way into the Quote-to-Cash process, how it will impact your business outcomes, and what steps you need to take to get started.

2 What is Quote-to-Cash?

Quote-to-Cash is the vital business process that spans a buyer's interest in a purchase to the company's realization of revenue. This means that Quote-to-Cash encompasses the entirety of a company's pricing, quoting, contracting, ordering, billing, renewals responsibilities and more.

The Functions of Quote-to-Cash:

- Product presentation: Navigating product portfolios, product selection and configuration (bundles, assemblies, components and options) and quote and order creation
- Pricing management: The full lifecycle of setting, executing and managing pricing, policies, and discounts
- E-Commerce: Delivering compelling digital buying experiences across sales channels
- Contract management: Generating, negotiating and signing contracts, as well as overseeing contract compliance
- Order management: Capturing and orchestrating orders across fulfillment processes
- Incentive management: Managing integrated rebates, promotions and sales compensation across direct sales teams, partner channels and E-Commerce websites
- Billing management: Billing, invoicing, and revenue recognition that supports different business models
- Renewals management: Identifying expiring contracts and effectively upselling, cross-selling and renewing business increasing recurring revenues and account penetration

These are all well-established functions, but often difficult to master as an integrated process, and even more challenging to innovate in order to achieve competitive advantage and optimal business outcomes. This is particularly hard for large global organizations with distributed workforces.

MACHINE LEARNING IN PRACTICE:

Cross-sell / Up-sell

What is it? Recommends products and services that your customers are most likely to purchase next, based on analysis of past purchases by similar customers.

Who benefits? This solution helps your sales reps increase deal size and increase account penetration based on machine learning insights.

An integrated, automated and agile Quote-to-Cash process increases productivity for businesses: eliminating manual tasks, keeping all stakeholders informed, collapsing cycle times, and lowering selling and service costs. Moreover, businesses attain superior visibility into how transactions are cultivated, further gaining a considerable edge in:

- Detecting market signals
- Pinpointing trends with buying and selling behaviors
- Understanding key dynamics that help accelerate or hinder sales

This combination of cohesion, visibility and insight empowers companies to quickly and easily respond to developments, and effect meaningful responses to market conditions. Simply put, Quote-to-Cash plays a central role in every company's effort to generate revenue, improve customer satisfaction, and realize profits.

The Vision for Apttus

In 2006, after having worked for 15 years for various software companies, Kirk Krappe, CEO of Apttus, made the following observations:

1. Automating a business process through a software package is good, but not sufficient to provide the business outcome that an enterprise seeks
2. The Quote-to-Cash process, arguably the most important business process of a company as it defines the revenue process, is "unautomated, manual and fraught with risks"
3. You can materially change the revenue of your company when pricing is first offered to a customer—at this moment, you must manage the process for maximum effect

Kirk recognized that there's an opportunity to change all of this. Guided by this vision, he, along with Neehar Giri and Kent Perkocha, founded Apttus, a Quote-to-Cash software provider.

3 What is Machine Learning?

The history of machine learning has proven that machines can learn and, thanks to Moore's Law (an observation by Gordon Moore of Intel in 1965 that computing power will double every 18 months), there's been a dramatic increase in the application of machine learning today. Machine learning parses data, recognizes patterns in that data, develops algorithms to achieve specific goals, and collects those algorithms into larger models that may be applied in other paradigms. Over time, as the machine acquires more data and recognizes more patterns, it strengthens the "learning" process and its data-driven recommendations.

Today, the average person, in multiple settings, has most likely encountered systems using machine learning.

- Every time Facebook proactively asks if you want to tag someone in a picture, the system is building on the data collected from a set of images where that person has been previously tagged, and with whom they have been previously grouped.

- Your purchasing history doesn't lie. When you used Amazon Prime to get overnight delivery on tropical fish food, size 2 diapers, and a carbon monoxide detector, the system added those tidbits to not only your profile alone, but also to all the other people out there with a rush delivery of fish food, diapers and home safety items to make sure that the next time you, or anyone "like" you, visit the site, machine learning algorithms serve up the products that you are most likely to purchase.

It's a Cat! Deep Learning Enables Computers to See

Instead of programming rules to recognize a cat, we feed millions of images of different kinds of cats into the computer and train it to recognize what a cat looks like—much like how a three-year-old child would learn from being exposed to millions or billions of images. In her Ted 2015 talk, Fei-Fei Li, Professor of Computer Science and Director of the Artificial Intelligence Lab at Stanford University explained how she did exactly that in 2009[1] with astonishing results. Machine learning lets the data (millions of images of cats) do the work by automatically figuring out the rules instead of a human manually entering rules. In so doing, machine learning is the way to make programming scalable—it is automating the automation.

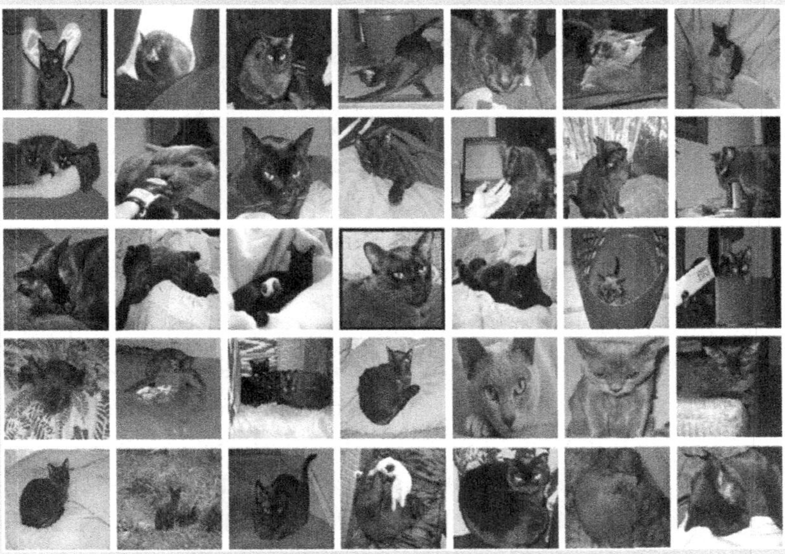

Figure 1. Through machine learning, a computer can be trained to recognize a cat even in different poses and backgrounds. Source: ImageNet.

---

[1] https://www.ted.com/talks/fei_fei_li_how_we_re_teaching_computers_to_understand_pictures

MACHINE LEARNING IN PRACTICE:
Pricing Intelligence

What is it? Recommendations for optimal price or discount levels for each quoted product based on past purchase history for the customer segment. This solution inspects related contracts to ensure that the contractual pricing terms are honored to provide the most appropriate renewal pricing.

Who benefits? Pricing intelligence helps your sales reps and sales operations deal desk in the quoting and negotiation process with guidance for initial, target, and walk-away price levels for every deal. With pricing intelligence, integrated with intelligent approval workflow you can now eliminate rogue discounting, curtail margin erosion and increase overall profitability.

- Did that scathing #doritos tweet you posted about a lack of crunchiness get a response? Machine learning can use language recognition to identify the sentiment of an online statement or review, and bring it to the attention of the social media team.
- That unsolicited offer from your bank to re-finance your home is likely based on more than just a consideration of your credit score (a machine learning process of its own) and payment history. Machine learning can consider all the mortgage data from your demographic, how often re-finance offers are accepted, what terms those successful offers contain, the age of the mortgage at the time that the offer was accepted, and any other factors that might affect your acceptance of an offer.

3.1 Machine Learning and Rule-Based Systems

Since the 1970s, rule-based systems—also known as expert- or knowledge-based systems have been used to capture and store human knowledge. Experts in the field regard these systems as a form of artificial intelligence where human knowledge is captured and "intelligently" dispensed.

Historically, these rule-based systems have been a key component of any Quote-to-Cash solution, where rules can be classified as follows: Constraint, Conditional, and Computation.

Constraint rules are declarative statements that express a restriction in the business process. For example, in a product configurator, a constraint rule can be written that expresses whenever product X is purchased, product Y must also be purchased, or no more than a quantity of five product Xs can be purchased by any one customer. Constraint rules can be mandatory or expressed as guidelines.

Conditional rules enable other actions to be taken if certain parameters or qualifications are achieved. If a shopping cart contains items over $100,000, then apply a 5% discount to the order.

Computation rules create new information, through a mathematical equation, by calculating the value of a business term, such as a shopping cart grand total. The total value of a shopping cart is the sum of all items plus the applicable taxes used for the location of the transaction.

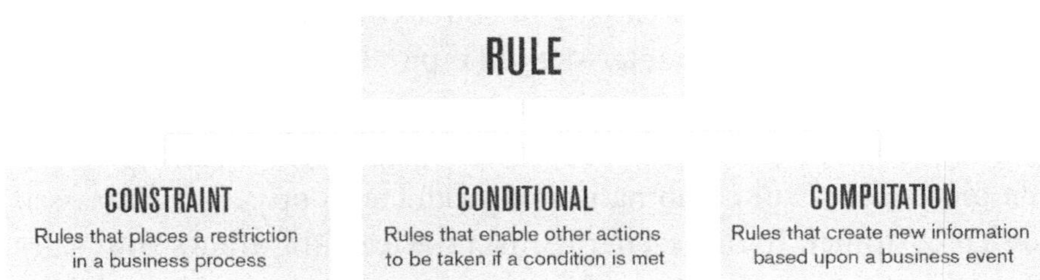

Figure 2. Types of business rules.

Rule-based systems work really well if every situation for which a decision must be made is known ahead of time. We see this with any "decision tree" diagram where answering yes to a question sends you down one path, and answering *no* sends you down another path. These systems embed knowledge or business policies in the process—making rules and decisions based on structured, finite, and constrained data.

But that's not the only kind of data we work with today. Today we work with vast amounts of data that is seemingly *infinite* and *unstructured*. Writing finite rules to deal with infinite data is impractical at best, and ineffective, or failing, at worst. Enter machine learning with its ability to derive knowledge. Machine learning algorithms can dynamically build, amend, and adapt insights as data is classified and categorized, enabling real-time decision making. Where rule-based systems are static, machine learning is dynamic.

Consider the potential of machine learning to aid with cross-sell/up-sell in the Configure-Price-Quote (CPQ) process. With traditional rule-based decision making, you could use historical human knowledge to write constraint rules as to which products to sell with another product. When you're dealing with dozens, or even a hundred products, this may be manageable. But when you increase the number of products to over 1,000 shop keeping units, or SKUs, and then add product data as it pertains to industry, geographic requirements, and company size, the complexity of human knowledge capture becomes dizzying and nearly impossible to execute against. Using machine learning with rule-based CPQ, you could employ an algorithm that examines customer characteristics like purchase history, size, location, purchased product combinations, and usage data - and utilize that information to identify a customer's propensity to buy, as well as the most successful combination of products to up-sell or cross-sell to that customer. In this scenario, machine learning allows your sales reps to maximize deal size and increase account penetration.

MACHINE LEARNING IN PRACTICE:
Quote Scoring

What is it? Predicts the probability of quote conversion based on the current sales stage of the quote. This solution also recommends next steps to improve the quality of the quote to increase conversion probability.

Who benefits? All available quote information is leveraged to identify similar historical quotes to recommend the next step so that all sales reps can sell like your top sales representatives.

The best practice is for rules based systems to work in concert with machine learning to produce a powerful combination of contextual knowledge and dynamically harvested intelligence. Together, they can drive the best outcome possible for your business.

Figure 3. Rules together with machine learning provides the best outcomes.

4 Data: Fueling Machine Learning

Machine learning is not a new concept; its etymology dates back to at least 1959, when Arthur Samuel, a professor at Stanford University, taught a computer to play checkers. But in the past decade we have seen an exponential uptick in the capabilities and achievements of machine learning. Why? Data. There's so much of it, and it's getting cheaper to gather, store and parse.

A recent study from IDC estimates that from 2013 to 2020 the digital universe will grow by a factor of 10, from 4.4 ZB[1] to 44 ZB. The study further submits that, while in 2013 only 22% of the information in the digital universe would be a candidate for analysis and less than 5% of that was actually analyzed, by 2020 the useful percentage could grow to more than 35%—mostly because of the growth of data from embedded systems.

We are in the middle of a perfect storm for data gathering:

1. the cost of computing power and storage is dropping as both become a commodity
2. the availability of cloud computing and cloud storage is rising with consumer demand
3. the pervasive use of personal and professional mobile devices
4. the proliferation of sensors and connected devices driving the Internet of Things (IoT), and
5. freely available open-source algorithms

---

[1] ZB (Zettabyte) = 1000^7 bytes = 10^{21} bytes

Figure 4. Drivers for machine learning now.

4.1 Harnessing Data to Find Meaning

In the old days, people took photos and put them in albums. When the albums were full they put the loose photos in shoe boxes. Machine learning allows us to gain an objective to our data gathering (previously a "throw it in the shoebox" exercise). It allows us to understand prior business phenomena, which may have been hiding under mountains of misunderstood or unstudied data.

The recording, aggregating and analyzing of data in the business arena is not a new concept. What's new is the growth and subsequent commoditization of enterprise data warehousing and business intelligence software. With the development of tools like Hadoop (an open source software framework for fast batch data processing across parallel servers) and a new class of databases known as NoSQL (delivering the ability to deal with relatively unstructured data) we are entering into an era that has been dubbed by some thought leaders as "Analytics 3.0." This is an era in which enterprises accept and embrace the notion that every device, shipment, and consumer leaves a trail, and that there is an opportunity to analyze those sets of data for the benefit of customers and markets.[2]

---

[1] http://www.emc.com/leadership/digital-universe/2014iview/executive-summary.htm

[2] https://hbr.org/2013/12/analytics-30

Data Scientists: Theorists, Dreamers, Tinkerers, Magicians

Data scientists, a label coined only a few years ago in 2008, are a new breed of high tech must-haves. As companies wrestle with unprecedented volumes and varieties of data, a data scientist brings the skills and the natural mathematical curiosity to put that data to good use. Without a data scientist managing and finding patterns and insights into your vast amount of untapped data, you're missing an enormous opportunity. By bringing meaning to large quantities of data, data scientists make analysis possible, empowering the feedback and optimization loop that is machine learning.

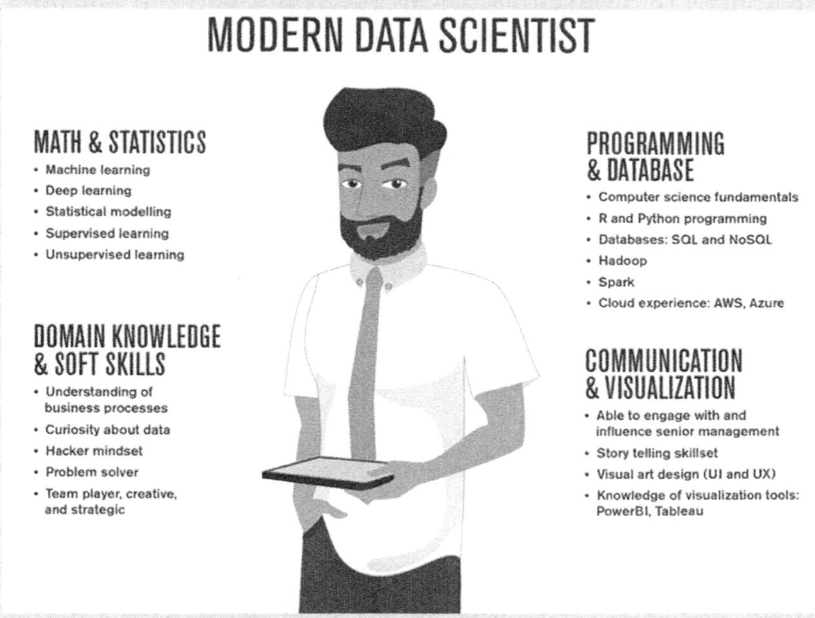

Figure 5. What it takes to be a modern-day data scientist.

Thomas Davenport writes in the Harvard Business Review that while a data scientists' most basic, universal skill is the ability to write code, the dominant trait among data scientists is an intense curiosity, one that drives a person to make discoveries while swimming in data.[1] Davenport further points out that some of the best and brightest data scientists are PhDs in esoteric fields like ecology and systems biology, like George Roumeliotis, the head of a data science team at Intuit, who holds a doctorate in astrophysics. When looking for that rare data scientist, it behooves us to focus on the "scientist" part of the term, not just the "data."

Sometimes the softer skills are the most difficult to find when looking for the ideal data scientist. The intense curiosity and problem solving that drives these individuals is wasted if the solutions cannot be communicated effectively, or built into a product that delivers the solutions to the fingertips of the users at the moment when it will have the most impact. The data scientist must possess the ability to work cross-functionally with product development, and the user interface and user experience teams to deliver the right data at the right time in the Quote-to-Cash process, or in any process, so that prescriptive insights are accessible and effective at the point of consumption by the user.

---

[1] https://hbr.org/2012/10/data-scientist-the-sexiest-job-of-the-21st-century

Historically, analytics in the enterprise were centered on reports and dashboards. These tools, created by business analysts for managers, were used to see what happened in the past. Their ability to improve business outcomes was limited by the amount of time (weeks or longer) required to create the report, the relatively narrow audience with access to the information, and the limits of value-added insight derived from the data.

Like milk or fresh produce, insights drawn from data have a shelf life and, if these insights are not consumed in a timely fashion, they will spoil. In their report on perishable insights, Mike Gualtieri and Rowan Curran of Forrester Research state, "If enterprises don't act on insights within a given time frame, they are no longer actionable; in other words, the opportunity to change the course of business outcomes will pass. The insights-to-action time frame—the perishability of the insights—varies depending on the type of insight and can range from microseconds to months and even years.[1]" For the Quote-to-Cash process, many of the insights are perishable within seconds or hours. For instance, information into how much to price a complex configured solution is often needed within a few hours because the prospect is waiting on the quote. The moral of this story is to not waste time and money on insights that you can't act on.

4.2 Using Data to Your Advantage

With the above points as a backdrop, let's take a look at the range of analytics available to the enterprise. Today, the vast majority of enterprises have needs for descriptive analytics, which are necessary for effective management. While descriptive analytics allow us to understand what has already occurred, and lays out relevant summaries and supporting data in formats that are easy to consume, it is not sufficient to accelerate business performance. It's only when machine learning is implemented—to facilitate predictive, prescriptive, or cognitive intelligence—that we can move from a reactive mode of business to one in which we are able to orchestrate the outcome of our efforts.

[1] Perishable Insights – Stop Wasting Money On Unactionable Analytics, Forrester Report, August 2016

The Intelligence Capability Pyramid

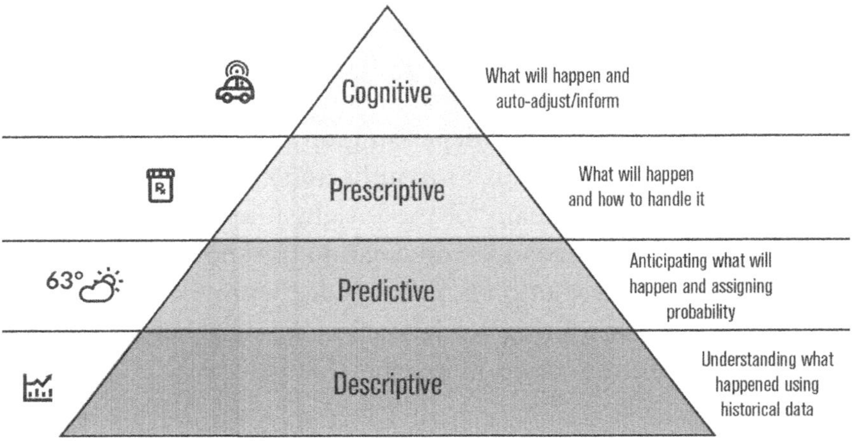

Figure 6. The Apttus Intelligence Capability Model summarizes the levels of analytics that enterprises need to scale to gain greater insights and to drive more profitable business outcomes

Descriptive Analytics

At its core, descriptive analytics is basic business intelligence. It is the foundation level that helps us understand what has already occurred, by laying out relevant summaries and supporting data in formats that are easy to consume both by end-user staff and management.

How can you use this type of analytics?

You can start with a big picture view of your metrics like bookings, revenue, recurring revenue, margins, or revenue velocity, and then drilldown for more granularity. For example, when it comes to pricing, your Quote-to-Cash system can analyze all of your sales contracts and provide a report on the pricing agreements by customer segments. Knowing this will allow you to decide which pricing to include in future deals of the same type.

Predictive Analytics
In a predictive analytics model, we use data from the past and assign probability to future outcomes. Predictive analytics helps users recognize patterns and detect meaningful trends. More significantly, you'll be able to generate projections of different developments for different time horizons, based on the output of the analysis.

How can you use this type of analytics?
Using the same example as above, your system can analyze successful deals, along with attributes mined from the marketing lead database. When this is matched with an internal product rank, you'll be able to predict which of your prospects will not only lead to a sale, but also which products they will likely buy. This allows you to prioritize specific accounts and plan accordingly.

Prescriptive Analytics
Once we know what is likely to occur and we have significant understanding of previous outcomes, we can prescribe a set of choices that the customer or user is most likely to agree to. For example, internet radio Pandora is powered by the Music Genome Project, a machine learning process that proactively serves up musical selections based on genre and artist preferences. Prescriptive analytics delivers granular insights and forecasts showcasing what is likely to occur, accompanied by relevant, system-driven recommendations on next best actions and tactics.

How can you use this type of analytics?
In the above example, if the Quote-to-Cash system is able to measure sales trends over an extended period of time, it will discern a spike in a particular product and provide recommendations to users at various stages of the Quote-to-Cash process. For instance, based on the insights, rather than the

normal 15,000 units you've committed for inventory, your system may tell you to allocate 25,000 extra units for a specific region. Now you're harnessing the power of machine learning to proactively react to the rapidly changing needs of the customer.

Machine Driven

Cognitive analytics exploits machine learning to refine trend and pattern analyses on an on-going, unsupervised, human-like basis in order to constantly evaluate processes and associate data. It also leads to an intelligent automation that initiates specific, suitable policies, actions and workflows that self-adjust and inform the user.

How can you use this type of analytics?

Imagine that your Quote-to-Cash system notices an influx of demand for a product in a certain region. Cognitive analytics allows the system to automatically adjust the price slightly to match demand, generating a greater profit. An opportunity that previously would have been missed or delayed has been fully realized and has positively affected the bottom line.

On their own, descriptive and predictive analytics are not sufficient to provide actionable suggestions. A subject matter expert with robust business acumen and domain expertise is required to interpret the insights and translate them into actions at the moment of consumption. This takes time, time that may spoil the insights if not promptly implemented.

Prescriptive and cognitive analytics build on and takes predictive analytics further by providing automated, actionable recommendations and implementing a feedback system that tracks the outcome produced by the action taken. These solutions consider possible outcomes based on different actions, and can therefore recommend the best course for any desired outcome.

5 Going Vertical with Machine Learning

Machine learning is making it possible for corporations to convert all the data they have been saving for years into a competitive advantage. Sales, marketing, and channel management teams can capitalize on machine learning's core strengths to determine which factors most influence buying behavior and which will accelerate bottom line revenue. Using machine learning, customer-facing teams can:

- Predict the propensity to purchase, and in which channel
- Make personalized recommendations to customers
- Optimize promotions for customers and partners
- Offer compensation and rebates to promote revenue growth in each sales channel
- Anticipate potential revenue and credit risks
- Forecast revenue and customer loyalty

And this is true across almost every industry. In a competitive landscape where challenges keep changing and data never stops flowing, machine learning is the tool that allows one enterprise to pull ahead of another. For example, in manufacturing, machine learning can be leveraged to optimize production and predict maintenance; in retail, it can be used for predictive inventory planning. Figure 7 illustrates what machine learning can do in different industry verticals.

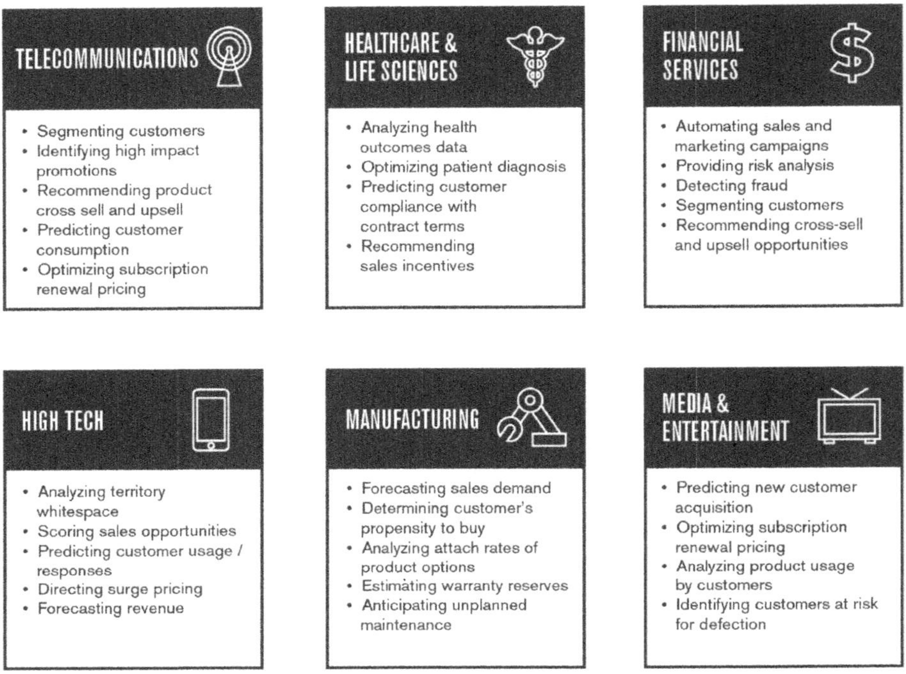

Figure 7. How machine learning is used across different verticals and industries.

Consider the effect machine learning can have on your Quote-to-Cash processes: Machine learning within Quote-to-Cash is leveraged for your business by using the entire sales and asset history, including customer characteristic, to find the patterns that link customers and the products they have purchased, and make intelligent predictions for products and services your customer or prospect may be likely to buy.

These predictions can be served to sales representatives or directly to E-Commerce customers as highly relevant options to consider.

Man-machine symbiosis: A lesson from Google's AlphaGo

Go is a game that originated in China more than 2,500 years ago. Two players start with a completely blank board and place black and white stones, one at a time, to surround territory. Once placed, stones do not move, and they're removed only when they're surrounded completely by the opponent's stones. The number of potential legal board positions is purportedly greater than the number of atoms in the universe. Teaching a machine to play Go is considered to be the Holy Grail of artificial intelligence.

So when Lee Sedol, the world's leading Go player, accepted Google's million-dollar DeepMind Challenge Match against artificial-intelligence program, AlphaGo in Seoul, Korea, for a best-three-out-of-five tournament, the general attitude was easy money for Lee.

Until play commenced. Over 70 million people watched in awe as AlphaGo shut Lee out in three consecutive games.

AlphaGo was not always unbeatable. Five months prior to Lee's losing series, Fan Hui, the then European champion, won 2 out of 5 games against the machine. But since that time, AlphaGo had nearly five months to improve, playing itself millions of times, never resting, incrementally revising its algorithms based on which sequences of play result in a higher win percentage.

But don't panic; this isn't HAL 9000, Megatron or Skynet. Harnessing the power of machine learning means optimizing our own tools to our best benefit.

Computer scientist Andy Salerno wisely offered this insight after Lee lost the Go challenge: "AlphaGo isn't a mysterious beast from some distant unknown planet. AlphaGo is us...AlphaGo is our incessant curiosity. AlphaGo is our drive to push ourselves beyond what we thought possible." And now Go players around the world are studying AlphaGo's moves in that tournament, learning from the machine and improving their gameplay.

We can do the same in the enterprise. Leveraging machine learning by harnessing its power in a complex, enterprise-wide business process like Quote-to-Cash, brings greater understanding, insight, and prescriptive intelligence to your business. It contributes to diverse strategic outcomes, reduces risk and increases sales efficiency. In the enterprise, we aren't in competition with machine learning—we are in competition with other enterprises, and the organization that most fully embraces and successfully exploits its data, by using machine learning to empower its people, wins.

6 Getting Started with Machine Learning

Deploying machine learning into a complex, enterprise-wide process —like Quote-to-Cash—begins with domain expertise, followed by time for the machine to prove its value and trustworthiness. Without Quote-to-Cash domain expertise, deploying machine learning through a platform will neither yield optimal results nor produce the well-thought out user experience that drives continued usage and adoption.

Quote-to-Cash affects teams across the enterprise, including sales, sales ops, marketing, legal, finance, operations, and IT. If any one of these participants decides that the predictive or prescriptive analytics are not the proper action to take at the proper time, they may ignore the guidance provided by the system. Even as the model is corrected and optimized and continues to improve, the user may hold doubts, defeating the purpose of integrating machine learning into the Quote-to-Cash process.

As you leverage machine learning, you must take the measured steps to identify and define the silos of cross-organizational data, build test cases, and ensure that stakeholders are invested in your organization's Quote-to-Cash processes. Identifying operational and financial metrics, quantifying the value of improving Quote-to-Cash performance, and carefully mapping out the details of each phase in the Quote-to-Cash process will result in better business outcomes, more invested stakeholders, and more trusting users.

6.1 What are Your Options?

Some Quote-to-Cash systems use an "opaque" box approach—one in which the deployment team cannot look under the hood to see how things like price optimization are calculated and prescribed. While some may see value in keeping that information confidential, it poses the challenge: if you can't see the model, how can you judge its accuracy or the trust logic behind it? In today's open source world, with freely available machine learning algorithms, maintaining proprietary models is legacy thinking. Contemporary thinking is to have a transparent box approach with prescriptive insights guiding the users.

Some B2B enterprises may entertain a do-it-yourself (DIY) approach to building machine learning into their Quote-to-Cash process simply because they employ data scientists (see Chapter 4.1 for what it takes to be a data scientist). In the case of deploying machine learning into Quote-to-Cash, it is extremely difficult to employ data scientists without the QTC domain expertise.

At Apttus, where Quote-to-Cash innovation is our business, we see models as one part of a prescriptive intelligence solution, along with data, presentation of guidance in the business process, trust in that guidance, and a consistent feedback loop. Our approach is to work with you, our customers and our partners, to leverage your expertise to develop the most effective prescriptive intelligence solution for QTC in the world. Our strength and focus is in delivering a solution that automates the creation and consumption of intelligence. The Apttus platform is used to seamlessly present trustworthy, proven, repeatable Quote-to-Cash recommendations to your users across the entire organization.

A typical QTC process can be characterized by a number of decisions that are made by the previously mentioned stakeholders within a company. Some of these decisions are common across many companies—which customer is a good target for our products and services? What should I offer to the customer? What is the right price to win the deal? Is it reasonable to assume we will have cash on hand for an upcoming expense? Is my customer at risk to move to another supplier?

There may be cases where decisions are unique to an industry. What price will increase the probability of winning the deal given the future utilization of my company's manufacturing plant? Or, specific to a company (for an online advertising solution provider) - what is the predicted lifetime value of a new customer?

These are all examples of decisions that are important to a company's revenue making process and questions that can be answered better by people who have insights based on analysis of historic data. And, they represent examples of cases where, over time, machine learning analysis will make the answers increasingly valuable.

7 The Future of Machine Learning in the Enterprise

Figure 8. Clippy, Microsoft's intelligent assistant, was bundled into Microsoft Office 97-2003 and was designed to help users navigate how to best use Office. Clippy, which popped up at inconvenient times, became more of an annoyance to users than a helper.

Call it the revenge of Clippy[1] or Clippy 2.0, but bots, also known as intelligent agents, are back and here to stay. There are many definitions for bots. Gartner's (an analyst research firm) formal definition is, "Bots are microservices or apps that can operate on other bots, apps or services in response to event triggers or user requests. They may invoke other services or applications, often emulating a user or app, or using an API to achieve the same effect. These requests can be initiated via conversational UIs or in response to a change in the state of a back-end application or database. Bots automate tasks based on predefined rules or via more sophisticated algorithms, which may involve machine learning."

[1] https://medium.com/@saranormous/clippy-s-revenge-39f7387f9aab#.ccy5zn7h8

Creator of Chatbot Beat 190,000 parking fines – Dubbed the Robin Hood of the Internet

UK born Stanford student, Joshua Browder, creator of DoNotPay, designed a chatbot lawyer to overturn thousands of parking tickets, saving over $5 million in fines.

Josh initially created his bot out of necessity, as he was getting parking tickets that he could no longer pay for, and as a way to show off his technical prowess to his friends. The bot would ask a few questions to understand the legal issue, the person's circumstances, and then commonly drafts a letter on behalf of that person.

That is just the start, Josh is turning his attention to homelessness and to help Syrian refugees gain asylum status in the UK through his automated bot. He is partnering with charities such as Centrepoint in the UK, that targets youth homelessness, to enable Internet and computer access to his bot.

A terse definition for our examination of bots in the enterprise comes from Phil Libin[1] where he defines bots as "interactive services that provide a conversational experience." The customer service use case is a quintessential example of where bots assist and guide customers through a difficult transaction, troubleshoot a problem, gather data, or make product recommendations. Remember the last time you contacted customer service to troubleshoot a non-delivered shipment or tried to determine why your credit card statement had an extra charge? If you are wondering what the weather is going to be like where you are at in a particular city, you can ask Siri on your iPhone. If you want to turn on music or turn off the lights in a room, you can ask Amazon's Alexa. Alexa can even order pizza for you, following your voice command.

When you consider the evolution of how humans interact with a computer, it started with punch cards, moved to a keyboard and then

[1] https://medium.com/@plibin/a-charge-of-bots-9ee33bb3b868#.4fnpd5uwn

to a mouse - to access programs or apps that are written to automate tasks. This may now seem outer-worldly with today's touch and swipe capabilities, but in the early days of computing, a user would access their software through a command line interface (CLI). As the user experience got more refined, we accessed our apps through clicks, drag and drop, the keyboard (keyboard shortcuts), or through touch. Bots are rewriting the rules for how we access our apps. CPQ, ERP, or CRM enterprise apps are just packaging for a series of tasks to be done. Imagine someday where the construct of an app is no longer constrained; we can type, speak, or think what we want to get done and the bot makes it happen across apps instead of moving through today's construct of a hierarchical graphical user interface (GUI).

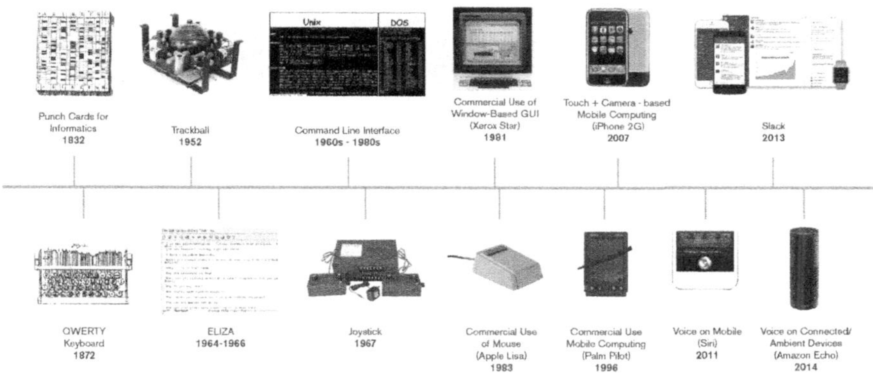

Figure 9. The evolution of human computer interfaces.

Imagine no further, as today's world is driven by social, mobile, virtual and conversational activities. This is the way people connect, and the way people—especially salespeople—work. But this is not the way apps are being built today. As a salesperson working within a Quote-to-Cash process, you are required to open your CRM instance, log in, and complete information about a meeting you have already had. The information entered is static. It's an old-school, hands-on, onerous process.

Looking to the future of the Quote-to-Cash process in the enterprise, machine learning is playing a pivotal role in moving the industry to reflect how we actually work in today's social, mobile, virtual, conversational world.

We can achieve this type of interaction through artificial intelligent agents or Bots, like Max from Apttus. AI agents are revolutionizing our interaction with apps, allowing sales people to track customer relationships and report sales using straightforward conversation with Max over the phone, Skype, or even HoloLens.

Figure 10. Different ways of interacting with intelligent chatbots.

How's that? Imagine you have just completed a sales call at Company J, and, as you're driving out of the parking lot, you get a phone call from Max, your personal, automated, Quote-to-Cash intelligent agent. Max is tied in to your Outlook calendar and sees that you had a meeting with Company J and, in a conversational tone, asks you how it went. As you tell her, she is making notes in your CRM, updating the record. She may note that you have an empty field for an executive sponsor and will ask, Who is your executive sponsor? When you respond with "John Ecks," she will complete the field, and let you know that John Ecks is tied to Jill Why and Jay Zee via LinkedIn. Using voice commands only, you can have Max reach out to Jill Why and Jay Zee to make further inroads for the deal. Max may ask if you have a close plan, and if you don't, Max may ask if you would like her to create one for you based on other deals that you have closed in the past.

Figure 11. Configuring and pricing products through a chatbot interface on a smart phone.

Further, you can tell Max that you're ready to create a quote for Company J. She will ask if there is another company for whom you have crafted a similar quote in the past. You might say, Yes, Company K is very similar. Max will create the quote, and ask if you would like to change the discount structure, or the product configuration. With machine learning, she can proactively suggest alternative terms and recommend pricing that have been successful in other quotes for similar customers. Once you're satisfied with the quote, she can send it to the customer on your behalf.

This is the reality of how we communicate today, and how we want to do business in the future.

> *Artificial intelligence, bots, and machine learning will be integrated into the fabric of our daily business workflow, augmenting our human capabilities, harnessing a seemingly infinite amount of data and presenting it in an instinctive, natural interaction.*

Like the Google search bar or Amazon product recommendations, these incredible AI capabilities will be ubiquitous, invisible and transparent in the app—you will no longer realize or notice it; it will be like breathing and you will use it every day.

8 Now is The Time

In the Quote-to-Cash process, machine learning is proving to be unequaled in delivering predictive and prescriptive insights, including defining which behaviors have the highest propensity to drive desired outcomes. If your organization is eager to compete and win more customers, applying machine learning to sales and marketing challenges is a proven competitive edge. Companies that have implemented machine learning with an automated Quote-to-Cash process have on average realized 105% larger deal sizes, 28% shorter sales cycles, and 26% more reps achieving quota.[1]

The Accenture Institute for High Performance recently completed a study[2] (presented in MIT Sloan Review) that found the following key takeaways:

- **At least 40% of companies surveyed are already using machine learning to improve sales and marketing performance.** Two out of five companies have already implemented machine learning based intelligence in sales and marketing.
- **38% credited machine learning for improvements in sales performance metrics.** Metrics the study tracked include new leads, upsells, and sales cycle times by a factor of two or more while another 41% created improvements by a factor of five or more.
- **76% say they are targeting higher sales growth with machine learning.** Gaining greater predictive accuracy by creating and

[1] https://www.docusign.com/blog/momentum-preview-unleashing-the-power-of-data-with-apttus-quote-to-cash/

[2] http://sloanreview.mit.edu/article/sales-gets-a-machine-learning-makeover/

optimizing propensity models to guide up-sell and cross-sell is where machine learning is making contributions to omni-channel selling strategies today.

Now is the time to break away from linear, task-oriented processes and embrace an outcome-focused, behavior-driven and machine-intelligent approach to Quote-to-Cash. Now is the time to step up and lead your company down the path to a smarter Quote-to-Cash execution. Do so and you'll drive maximum business results.

Appendix A
Development Timeline of Artificial Intelligence

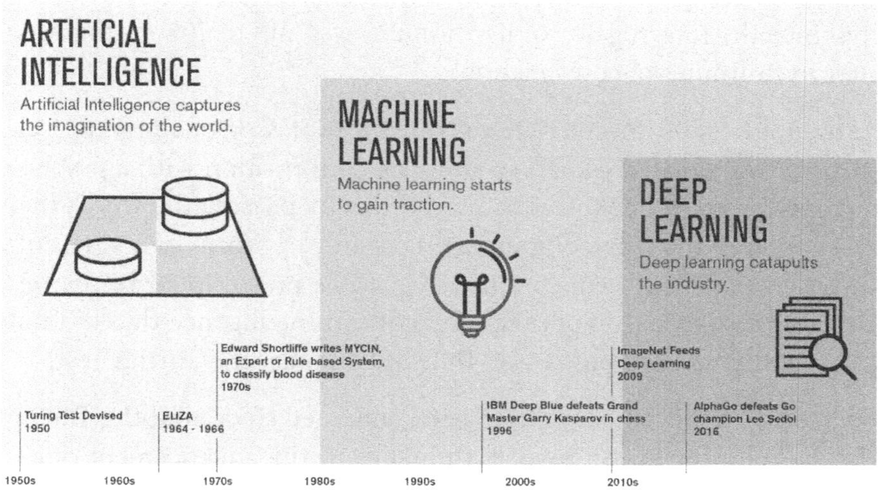

Figure 12. Timeline of Artificial Intelligence development.

Artificial intelligence is an umbrella term under which machine learning and deep learning fit snugly (see Figure 1). Coined in the 1950s, AI is a broad reference that includes a set of methods, algorithms and technologies that make the underlying software seem capable of exhibiting behavior or intelligence which is indistinguishable from a human.

The most widely-known AI experiment is also one of the first: The Turing Test developed by Alan Turing in 1950 at the University of Manchester. In his seminal paper, *Computing Machinery and Intelligence,*[1] Turing considers the question, "Can machines think?"

[1] https://en.wikipedia.org/wiki/Computing_Machinery_and_Intelligence

Simply put, the Turing Test is a criterion to determine whether a computer has human-like intelligence. Dr. Turing postulated that a human interrogator would be unable to distinguish between the responses of a human and a computer over a text-only channel using only a screen and a keyboard (taking vocal responses out of the equation). His hypothesis was that, judging by the answers given from either the computer or the human interviewee, the machine could convince the interrogator of its "human-ness" up to 70% of the time after five minutes of conversation.

In the mid-1960s, Joseph Weizenbaum, a professor of computer science at MIT, created ELIZA to simulate conversation with a psychotherapist[1]. Professor Weizenbaum did this by parsing the words that users entered into the computer (this is an early form of a ChatBot) and then matched them to a list of possible pre-scripted responses. This was a rule based approach to artificial intelligence that did not have contextual awareness, and thus did not pass the Turing Test.

Pioneers like Weizenbaum have been emulated since, and the Turing Test has continued to prove its influence in the advancement of artificial intelligence with annual competitions like the Loebner Prize, that awards prizes to the "chatterbot" considered by the judges to be the most human-like.

In the 1980s computer scientists elevated the question from "Can machines think?" to "Can machines learn?" This was the advent of the field of machine learning, with its feedback loop for self-perpetuating improvements.

---

[1] https://en.wikipedia.org/wiki/Person-centered_therapy

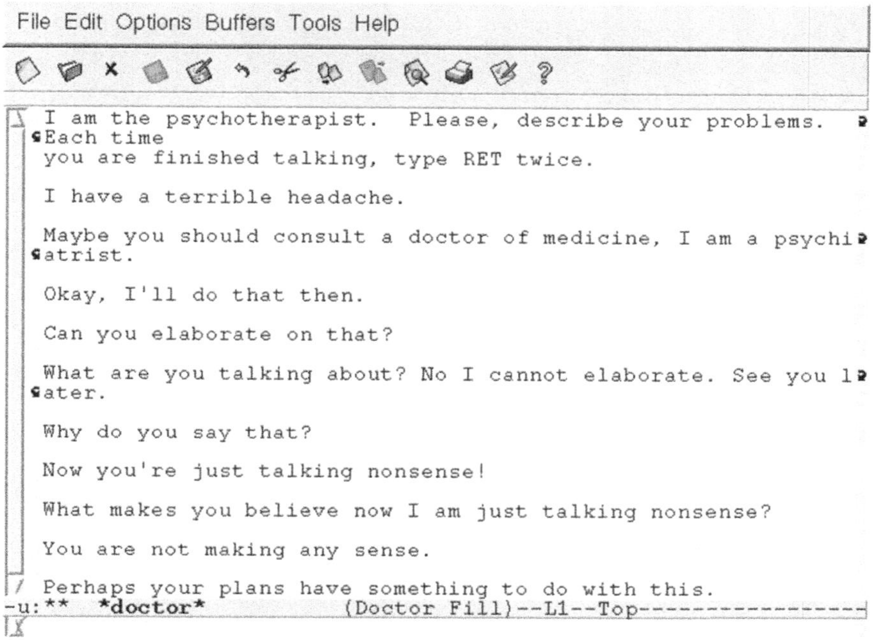

Figure 13. ELIZA is simulating a psychotherapist—an early ChatBot.

Since 2010, scientist have gone even further in machine learning to develop the field of deep learning, which takes its inspiration from how the human brain works. The idea is to simulate the large array of the brain's neurons in an artificial neural network (ANN). With the explosion of computing power, computer scientists can now add more and more layers of these ANN (depth) to the learning process. This technique has applications to computer vision, natural language processing, robotics and more.

Appendix B
The Basics of Machine Learning

Most people will not work on their own car engine, but a basic understanding of what is underneath the hood makes us more informed, lest we undertake unnecessary car repairs. As machine learning drives more and more of your B2B enterprise processes, it is worth taking a peek under the hood to see what is happening.

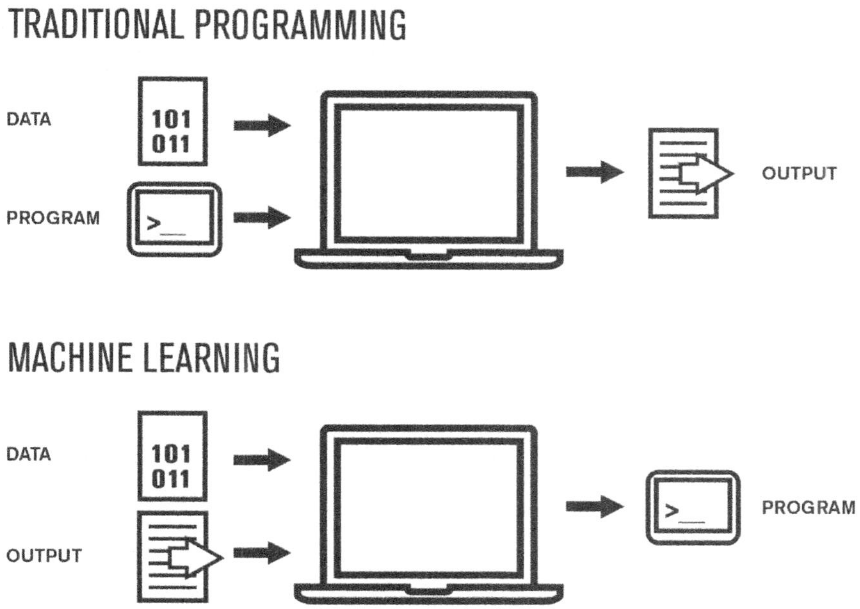

Figure 14. Traditional programming versus machine learning paradigm.

Programming By Input-Output Example

With enough data and computing power, machines can learn to predict the outcome of new data as it is introduced and prescribe the next course of action to achieve the desired goal. Thomas Dietterich, Oregon State University professor and president of the Association for the Advancement of Artificial Intelligence calls this "'programming by input-output examples' rather than by coding."

In traditional programming, a program, or a set of machine-readable instructions for the computer to automate tasks, is written to use data, which has been entered into the computer, to deliver output. Consider the example of programming the computer to recognize a cat. You can write a set of rules to approximate what a cat looks like: whiskers, pointy ears, fur, and a tail. Needless to say, a human programming the rules for a computer to recognize all the variations of a cat is tedious and futile. If the cat is partially hidden behind a tree or a rock, the program will not properly recognize the cat. Machine learning significantly alters this paradigm. The data and the desired output are reverse-engineered by the computer to produce a new program that can effectively predict the output based on the supplied input data.

Supervised and Unsupervised Learning

There are two general categories of machine learning—supervised learning and unsupervised learning.

Supervised Learning

In supervised learning, the machine learning algorithm has an explicit goal to reach, and is tasked to understand the data and the decision paths taken to achieve that goal. As more data is introduced into the system, the machine learning algorithm can recognize and optimize the path or paths to that goal, and further can begin to predict that certain data points will or will not result in that desired goal.

Depending on the type of output desired, supervised learning is broken down into three sub-categories:

Identification
Identification is simply understanding where a data point fits with all the other data points available. Using pattern recognition and set of finite outcomes, identification is like a game of twenty questions where, with the right questions, the output is a correct identification of any item in the universe.

Classification
Classification further parses identified data points into a finite result—is this data point one thing or the other, A or B? When you wrote that scathing #doritos tweet in the example above, classification algorithms identified the output with a finite set of options: positive, negative or neutral. "Hey @ doritos your newest flavor isn't crunchy like the original #wheresthecrunch," is identified as negative. Whereas when your neighbor tweeted: "Can't wait to serve the new @doritos flavor at my party! #partytime #dontinvitetheneighbors," the statement is identified as positive.

Regression
When the desired output is a continuous number, like a probability or a score, a regression algorithm is in play. Credit scores are an example of an outcome based on regression algorithms, taking into account your credit history, purchasing history, and the likelihood of debt payment.

Unsupervised Learning
In unsupervised learning the machine learning algorithm has no explicit output goal. Typically, unsupervised learning is used to ascertain an implicit, hidden structure and to uncover relationships between data points.

Clustering

Clustering algorithms are a subcategory of unsupervised machine learning, where the machine is tasked to find similarities in data points or groups of data. For example, Amazon uses clustering to offer up items that you will probably find useful. Netflix uses that same approach to suggest shows or movies that you are likely to enjoy. Because clustering has no determined output goals, the data can be parsed in multiple ways and presented to the user in what the algorithm has determined to be the most influential combination.

Appendix C
Data Science Meets Behavioral Science

In the United States alone, thirty-eight million people[1] start their day by eagerly fastening a device to their wrist that is not worn for the purpose of fashion or keeping time. It is a fitness tracker and these little gadgets have swept the nation. Why? Because we love having instant access to our performance, activities and goals. We enjoy tracking our progress throughout the day. We are addicted to the gratifying notifications of success, and the social aspects of competing with friends, family members and coworkers. The fitness tracker market has achieved tremendous success by providing us with relevant data and motivating incentives. These trackers are successfully inspiring the world to be more active by leveraging principles from both data science and behavioral science.

For centuries, traditional economic theory dictated that humans make logical, self-interested decisions, always choosing the most favorable conditions. However, reality often demonstrates otherwise. Every January, how many people do you know say that they want to resolve to save more, spend less, eat better, or exercise more? These admirable goals are often proclaimed with the best of intentions, but are rarely achieved. If people were purely logical, we would all be the healthiest versions of ourselves. However, the truth is that humans are not 100% rational; we are emotional creatures that are not always predictable. Behavioral economics evolved from this recognition of human irrationality. Behavioral economics is a method of economic analysis that applies psychological insights into human behavior to

[1] https://www.statista.com/statistics/413225/wearables-worldwide-unit-sales-forecast/

explain economic decision-making.[1] Essentially, it is the intersection between economics and behavioral psychology. Behavioral economics helps us understand why only one-third of Americans[2] floss daily, why most people's expensive home treadmills turn into overpriced coat racks, and why motivating humans is more complicated than ever before.

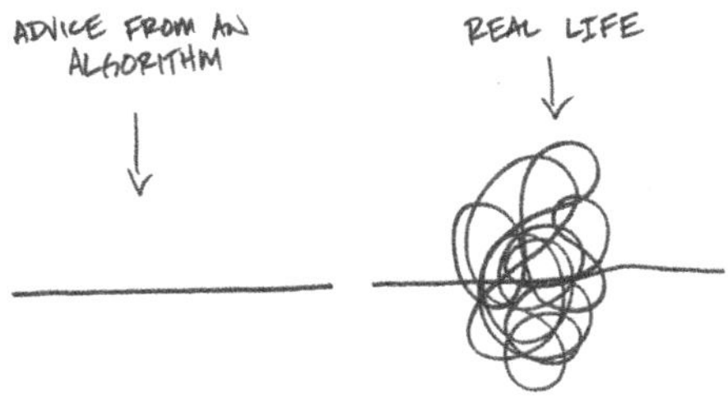

Figure 15. Traditional economic theory does not address human irrationality. https://www.behaviorgap.com/blog/

Human behavior can be seen as the byproduct of millions of years of evolution. With a nature forged from hunger, anxiety and fear, it is no wonder the behaviors of modern man can often be irrational—driven by forces like peer pressure, availability bias and emotional exhaustion. To change human behavior, we must embrace our human nature, instead of fight it. And one of the most powerful tools to help enable change is data.

Data science is the discipline that allows us to analyze the unseen—and with machine learning, it allows us to look at large sets of data and surface patterns, identifying when past performance is indicative of future results. For instance, it lets us forecast what products are most likely to be sold and which customers are most likely to buy. But what if you not only want to understand potential outcomes, what if you want to completely change outcomes, and more specifically, what if you want to change the way in which people behave? Behavioral

[1] https://en.oxforddictionaries.com/definition/us/behavioral_economics

[2] http://www.usnews.com/news/articles/2016-05-02/how-many-americans-floss-their-teeth

economics tells us that to make a fundamental change in behavior that will affect the long-term outcome of a process, we must insert an inflection point. What is the best method to create an inflection point or get someone to do something they would not ordinarily do? Incentives.

As an example, you are a sales rep and two years ago your revenue was $1M. Last year it was $1.1M, and this year you expect $1.2M in sales. The trend is clear, and your growth has been linear and predictable. However, there is a change in company leadership and your management has increased your quota to $2M for next year. What is going to motivate you to almost double your revenues? The difference between expectations ($2M) and reality ($1.2M) is often referred to as the "behavioral gap" (see chart below). When the behavioral gap is significant, an inflection point is needed to close that gap. The right incentive can initiate an inflection point and influence a change in behavior. Perhaps that incentive is an added bonus, President's Club eligibility, a promotion, etc.

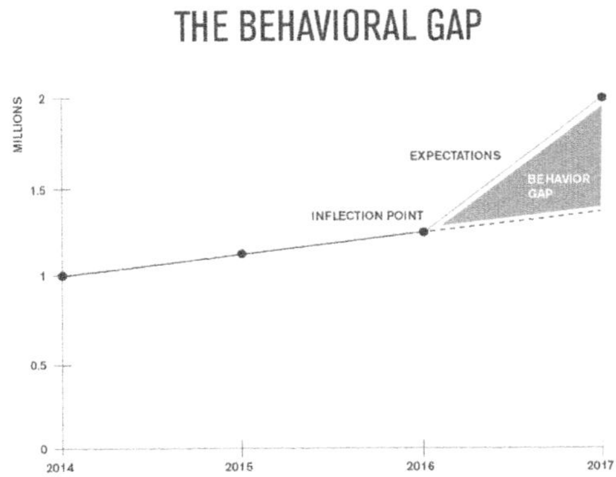

Figure 16. The behavior gap depicted above represents the difference between raised expectations (management increasing quota) and the trajectory of current sales performance.

In the U.S., studies from *Harvard Business Review* and various industry publications posit that companies spend over **one trillion dollars** annually on incentives. That number is four times the money spent on advertising in the US annually. What that means is that, as a nation, we are deeply invested in incenting people to act in ways that are somewhat contrary to how they would normally act, if left to their own devices. Incentives appear in many forms such as commissions and bonuses for sales personnel and channel sellers, rebate payments and marketing incentives for partners and customers, and promotions, discounts and coupons for end consumers. Deloitte University Press published a report[1] stating that when it comes to the relationship between data science and behavioral science, "it is reasonable to anticipate better results when the two approaches are treated as complementary and applied in tandem. Behavioral science principles should be part of the data scientist's toolkit, and vice versa."

---

[1] http://dupress.deloitte.com/dup-us-en/deloitte-review/issue-16/behavioral-economics-predictive-analytics.html

Data scientists work with product and sales teams, employing data and patterns to manage incentive programs. Using forecast modeling and behavior mechanics, teams can plot out the path from one goal to the next and analyze and implement proper incentives. As an example, let's say your company is a furniture manufacturer that uses a CPQ tool to manage its complex quoting and pricing processes. One of the major reasons your company invested in the CPQ solution was to curb chronic, costly discounting by the sales team. You are a new sales rep using CPQ to build a quote. What if, mid-quote, your system alerts you that the discount you entered, while within the approved range, may not be ideal. Machine learning ran in the background and identified a different discount used by the top 10% of reps that has had more success. Additionally, you learn that if you choose the prescribed discount, you will earn 40% more commission! Talk about a relevant incentive, based on powerful data.

In a real-world implementation, one Quote-to-Cash customer—let's call them Company X—who links websites with advertisers, needed to be able to better forecast the potential revenue for each deal. The nature of the business does not allow Company X to recognize revenue until a user clicks on an ad. They used Apttus to harness machine learning to understand past behavior, used behavioral science to influence future behavior, and implemented A/B testing (comparing two versions of a web page to see which performs better) on incentive effectiveness programs. The A/B testing data allowed Company X to understand the effectiveness of certain incentives to guide customer behavior.

When applied together, data science and behavioral economics provide powerful business results by collecting relevant, timely insight and defining incentives that align human behaviors with organizational goals.

Appendix D Reimagining the Role of the B2B Salesperson

Our personal purchasing expectations have been shaped by our experience in the consumer (B2C) marketplace—especially in regards to E-Commerce. 96% of Americans are shopping online these days, and even consumers who are purchasing items in a store are reading through online reviews and ratings before many purchases are made[1]. That carries through to our professional purchasing expectations. There's a driving demand for the B2B marketplace to catch up to the transparency, convenience, and ease of the B2C marketplace.

B2B companies that provide frictionless consumer experience capture more market share. In his report Death of a (B2B) Salesman, Forrester analyst Andy Hoar writes that B2B buyers want to self-educate, instead of engaging a sales representative about products and services, and a vast majority of enterprise buyers (93%) find buying online more convenient than buying from a salesperson. Increasingly, enterprise buyers have decided what they want to purchase before they engage with a salesperson, if they ever do.[2]

This could be a boon for the B2B salesperson—the majority of the sales process has already occurred by the time the salesperson gets involved. But the old-school sales guy working the phones and taking clients out for a round of golf is going to be left in the dust. In his report, Hoar writes that there are four new B2B buyer, and corresponding seller, archetypes, and almost all of those seller archetypes

---

[1] https://www.bigcommerce.com/blog/omni-channel-retail/

[2] https://www.forrester.com/report/Death+Of+A+B2B+Salesman/-/E-RES122288

are being threatened by the advancement of machine learning and its influence over E-Commerce capabilities.

Buyer: "Serve Me" | Seller: Order Taker

The buyer is purchasing a low-priced, minimally complex product, already knows what he or she wants, and is looking for the most frictionless way to get it. Hoar postulates that the order taker role is vulnerable and will be highly displaced across almost all industries by 2020, replaced with self-serve online E-Commerce.

Buyer: "Show Me" | Seller: Explainer

This buyer is purchasing a highly complex product or service. This individual has already done some product research, but because of the complexity of the product, he or she wants to have a direct relationship with the salesperson. The salesperson's role is to provide a supplementary level of information to the buyer. The explainer archetype will decline as well, by as much as 25% by 2020 according to Hoar, as companies productize content through FAQs, online product "university" portals, videos, webinars, and leverage syndicated user-generated content.

Buyer: "Guide Me" | Seller: Navigator

The buyer is already educated and "decided" but needs some guidance as they consider a purchase, perhaps because of the more complex environment with multiple internal stakeholders or challenging budget considerations. Here, the salesperson helps the buyer get to a solution that suitably addresses internal concerns. According to Hoar, the navigator archetype will decline as much as 15% by 2020, but their role can be up leveled to that of the consultant.

Buyer: "Enlighten Me" | Seller: Consultant

This buyer is considering a highly complex product or service, and is also dealing with complex internal hurdles. The consultant archetype is the role model within B2B selling, and these sales jobs are expected to grow by 10% by 2020. This is the only archetype that Hoar considers not to be under threat from advances in technology or B2B buying behavior, because this salesperson is the most adroit in relationship and solution selling, integrating themselves most effectively into the customers' business, and evolving from seller to trusted advisor.

As the B2B marketplace continues its quick-time march toward consumer-like experiences, the burden is placed on the salesperson to add value and reimagine his or her role. Machine learning is allowing E-Commerce transactions to become far more sophisticated and efficient. Does this mean that the future role of the salesperson is to simply bring the human touch to the sales process? And if so, how long is that salesperson integral to the customer's relationship with the company, and how must the salesperson truly affect the customer lifetime value?

Machine Learning Is Elevating the B2B Salesperson

As they say, the death of the B2B salesman is greatly exaggerated. Machine learning is the secret weapon that allows the B2B salesperson to be an agile, sophisticated and trusted advisor to his or her customers.

This is where leveraging machine learning within the Quote-to-Cash process becomes critical. When you consider that machine learning can be used to harvest additional data points from the CRM system and/or the marketing automation system within the Quote-to-Cash process, the ability to predict the success of each interaction increases. This automated feedback and optimization loop within machine learning is called **predictive intelligence.**

So what's the big deal? We know that in an E-Commerce transaction, it's not too difficult to predict that when the previous 8,200 shoppers bought that 4-D-cell Maglite that you just put in your online shopping cart, 72% of them also bought universal mounting brackets, so you probably will too. But what about those more complex B2B sales situations—the ones in which your customer isn't even sure what to put in his or her cart in the first place? That's when we turn our machine learning algorithms to **prescriptive intelligence.**

When a B2B salesperson is selling a highly complex product or service, leveraging machine learning within the Quote-to-Cash process means that you can harness data from the CRM system to develop a demographic of the buyer's company (e.g., companies with a similar size, revenue, industry, and/or competitive landscape) and make targeted recommendations to the buyer. Conversely, when powered by machine learning, prescriptive analytics can recognize when a potential customer is more likely to not result in a sale. Guided by machine learning, the salesperson now knows to walk away from a prospect with a low potential of actual revenue, focusing his or her efforts on a more promising opportunity.

Prescriptive analytics looks at the customer's demographics, as well as the details of the sales processes that have historically resulted in guiding that demographic to a successful outcome (i.e., closed deals and recognized revenue). Once the seller has been guided towards providing the right options for his or her prospective buyer, the salesperson can further rely on prescriptive analytics to recognize revenue maximizing opportunities, like high propensity up-sells and cross sells.

Next, the salesperson gains even more momentum with prescriptive pricing—a customized pricing structure with discounts that have most often resulted in a closed deal with this type of buyer. Prescriptive pricing uses machine learning to advise the salesperson of the right price to offer at this point in time, within company-approved parameters, which will give him or her the best chance of winning the deal.

Research and discovery, based on machine driven insights, allow the B2B salesperson to achieve a higher level of sophistication in the sales conversation. By embracing machine learning and evolving into the "consultant" archetype, the seller becomes the ultimate trusted advisor.

Made in the USA
Monee, IL
09 August 2021

75304300R00036